Graph Paper Notebook

Quad Grid 5 X 5

Size 8.5" X 11"

Taylor Madison

Graph Paper Notebook 1

If You Love Our Books
As Much As We Love Bringing Them To You,
Would You Consider Leaving a Review?
It Could Really Help Us! Thank You!

Taylor Madison

Be

Made in the USA
Monee, IL
08 November 2026

67950731R00063